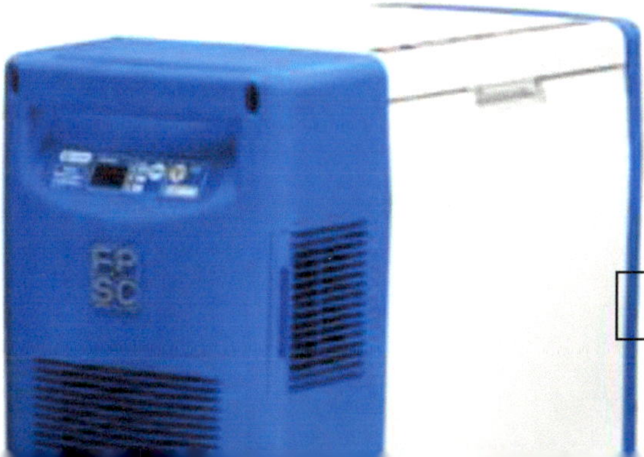

The Outreach Cooler

Table of Contents

I. Executive summary

The Cool Mile division of Global Cooling markets and sells the world's only Free Piston Stirling Cooler (FSPC) vaccine refrigerator - the *Outreach Cooler.* Based on patented stirling technology owned by Global Cooling, the refrigerator maintains an interior temperature of 4° C under a range of ambient temperatures as high as 43°C. The key differentiating features are:

- Lightweight - 32 lbs vs. 90 lbs for a comparable unit
- Portable - runs on a 12V charger as well as solar power and grid electricity
- Highly efficient - uses 80% less power than similar sized compression refrigerators
- Reliable, Durable, Long life span - 30 years vs. 10 years
- Mixed Energy Inputs – Runs efficiently on 12 volt, solar energy and grid electricity
- Environmentally friendly - cools using neutral helium versus CFCs or HFCs
- Price Competitive - $2000 USD

These features make the *Outreach Cooler* the perfect refrigerator for storage and delivery of vaccines for immunization in rural health centers, especially in low and middle-income countries. Most vaccines must be kept at 5°C or below to remain viable. Currently, over 50% of vaccine doses are wasted due to exposure to heat or freezing. (PATH, 2008)[1]

The vaccine refrigerators in place today are largely responsible for this waste. They are unreliable and are expensive to operate due to their high consumption of energy. In Uganda, 30% of the vaccine refrigerators were out of service due to breakdowns or waiting for spare parts during an audit in 2007. (PATH, 2007) Moreover, health centers in low and middle-income countries experience frequent "brown outs" and "blackouts" in electricity during which the vaccine refrigerators lose power. To ensure that the refrigerator will be functioning as needed, many existing units use propane or kerosene as a primary or alternative power source. In addition to being expensive, propane and kerosene can be difficult to obtain and to maintain a constant supply. (Village Reach, 2008)

The *Outreach Cooler* differs from the competition in that it will run for 30 years, maintenance is as simple as cleaning the filter, it uses less energy whether solar or grid electricity and it can run well under "brown out" conditions that compression refrigerators cannot. It meets the needs of the customer to have a vaccine refrigerator that is cheaper to run, works longer and is more reliable than competitors.

---

[1] PATH is the leading innovator in creating solutions to break longstanding cycles of poor health. Cool Mile is privileged to have the benefit of their help and guidance on this project. For more about PATH and our other partners, see Section IX. Partners

Immunization is one of the most effective and inexpensive ways to increase health and wellbeing. (Abt Associates Inc, 2000) Cool Mile's customers are Ministries of Health (MOHs) and private hospitals and clinics in low and middle-income countries that are focused on delivering vaccines as part of their health services. All of these customers have the need to increase the "up time" and reliability of vaccine refrigerators and decrease operating costs – i.e. fuel and maintenance. (WHO, 2005; PATH, 2008)

The Cool Mile division is a profit center within Global Cooling. Gross margin in 2011 is expected to be $650K on a volume of nearly 800 units, generating top line revenues of $1.4M. The division is cash flow positive from the beginning and will breakeven in year 2, recovering the initial investment at that time. In year 5, EBITDA is $1.2M on a volume of 1,900 units.

The next step is modifying the existing product to pass the pre-qualification system for approval by the World Health Organization (WHO). This is the "seal of approval" for Ministries of Health and others that the product is vaccine quality. This is in process and will be complete in October, at which time Cool Mile will begin an in country test of the *Outreach Cooler* to gather data and customer feedback about how the product performs under conditions in the field. Total cost for these two steps is $800,000.

The combination of positive cash flow, environmental benefits and the social good of increasing immunizations makes the *Outreach Cooler* a truly triple bottom line solution to vaccine refrigeration. The innovative stirling engine technology is patented by Global Cooling. The long term goal of the Cool Mile division is to establish proof of concept in form of the Outreach Cooler and then sell or license the technology to a larger player. This will position Global Cooling and Cool Mile to continue to innovate in the area of high end cooling solutions and deliver on their mission to change the way the world cools.

II. "The Outreach Cooler" Product Overview

Outreach Cooler Product Specifications			
Technology	Stirling Engine	**Number of Baskets**	2
Gross Capacity (liters)	25	**Power Source**	DC12V
Vaccine Storage Capacity @ +43 degrees	9.2	**Energy Consumption (Watts)**	48
Vaccine Storage Capacity @ +32 degrees	9.2	**Hold over time during power failure (hours)**	4.5
Outside Dimensions	-	**WHO Specification ID**	E.4
Width x Depth x Height (mm)	695 x 350 x 460	**Cooling Refrigerant**	Helium

The Outreach Cooler is the size of a typical recreational ice chest. It has a small cooling engine and electronic temperature control panel on the side. It weighs 15kg/32lbs and can support a health center with a catchment area of 5,000 people2, depending on supply intervals. The price is $2,000 USD, which is comparable to compression and absorption units of a similar size (see A. Current Competitors). It can be used as a stationary refrigeration unit in a health center for vaccine storage or as a transport unit during outreach campaigns where a healthcare worker might be away from the clinic administering vaccines in remote areas. It runs on electricity, either from the grid, a 12-volt system or from solar panels. The Free Piston Stirling Cooler (FPSC) technology runs well on variable energy input levels. Compression or absorption refrigerators cannot handle variability in energy input, which leads to their breakdown. This is a distinct advantage of the Outreach Cooler in "dirty" grid scenarios or running on solar power.

Energy inputs are the largest part of a country's operating expenses after salaries. Uganda spent $1.7 million dollars on gas for refrigerators in four years from 2006-2010. (WHO, n.d.) The *Outreach Cooler* will lower operating costs by running well on solar which has very low operating costs, or running more efficiently on

---

[2] Cool Mile estimate based on Partners in Health catchment information and PATH estimates of population ratios to number of health center averages in low and middle income countries.

electricity. Cool Mile's test shows that the unit uses up to 80% less energy than refrigerators in place today.

Based on an existing product from Global Cooling, the *Outreach Cooler* has been modified to meet the stringent requirements of the World Health Organization PQS system to be designated as a vaccine quality refrigerator. The modifications include:

1) Adapting the power supply to connect directly to photo voltaic solar panels;
2) Changes in electronic controls to handle multiple power inputs (solar, grid, 12-volt);
3) Adding phase change material to increase the interior temperature "hold over" time during power outage.

Health centers in rural areas or outreach programs use this product to keep vaccines cool during storage and in transit to remote locations. As the moving parts are internal to the engine, the maintenance is minimal and can be done by non-technical health center staff. All that is needed is to slide out the filter from the side of the unit and clear any lint or dirt. No tools or expertise is required. Because the unit is portable, the need to transfer vaccines to a passive cooler for outreach is eliminated, reducing exposure to heat and exposure to freezing in passive coolers, cooled by ice packs.

Product Sustainability

The *Outreach Cooler* is similar to its competitors in that it is has a plastic body, foam insulation and phase change material made from a paraffin derivative. The cooling engine is very different however, running on helium, which is a neutral gas with no environmental impacts. (Helmenstine, n.d.) Existing products use CFCs and HFCs as refrigerants that have proven to damage the ozone layer. (CIESIN, n.d.) The engine also runs on considerably less energy, making it more sustainable from both a cost and environmental standpoint. The expected 30-year life span of the unit reduces the need for replacement, using fewer resources than those that need replacement every 7-10 years. The Outreach Cooler is far superior to competitive products in lowering its impact on the ecosystem.

III. Immunization and the Right to Live a Full Life

Vaccines, which create immunity to disease, are administered around the world on the principal that it is easier and more effective to prevent disease than to treat it. Suffering, disability and death are avoided, the contagion is reduced and health care services and dollars can be used elsewhere. The World Health Organization (WHO) estimates that 2 million deaths were averted due to immunization in 2002. In fact, immunization can virtually eradicate disease. Cases of smallpox and polio have been reduced by 99% and measles by 68% since the introduction of focused immunization campaigns. (WHO, n.d.)

Yet, immunization is not routine everywhere in the world, particularly in remote locations, urban slums, and border regions. Poverty is almost synonymous with lack of access to healthcare and immunization. Diseases like measles, mumps, rubella and tetanus, which are almost nonexistent in developed countries, are common in low and middle-income countries. These preventable diseases can impact society as a whole by lowering time spent in school due to illness, time spent away from work to care for sick family members, long term disability and an increasing percentage of income spent on medicine and healthcare services. (Bloom, Canning, Weston, 2005)

The ability to live life to the fullest of human capacity starts with health. Routine immunization for everyone is a question of equity. The *Outreach Cooler* enables health care systems to overcome one of the critical barriers to immunization – storage of vaccines in remote areas, where electricity is not a given, and health clinics serve thousands of people over a large area. By providing a product that helps reach more people with vaccines, the Cool Mile division is working to level the playing field in health.

The Cool Mile division takes sustainability into consideration in its product specifications, operations, marketing and sales. With a firm commitment to minimizing the environmental and social impacts of marketing, selling, distributing, and servicing the cooler, the Cool Mile division has an ethic of continuous improvement in the area of sustainability. For details of these considerations please refer to the individual sections on these topics.

IV. The customers and the market

A. Market Forces

There are three interdependent forces converging on the vaccine refrigeration market – increased funding for immunization, increased focus on operating expenses due to failures of existing cold chain equipment, and the expected delivery of new, expensive vaccines. Each of these factors has created momentum in the NGO and international community to solve the existing issues in the cold chain.

The primary issue is waste. Vaccines must be kept at a

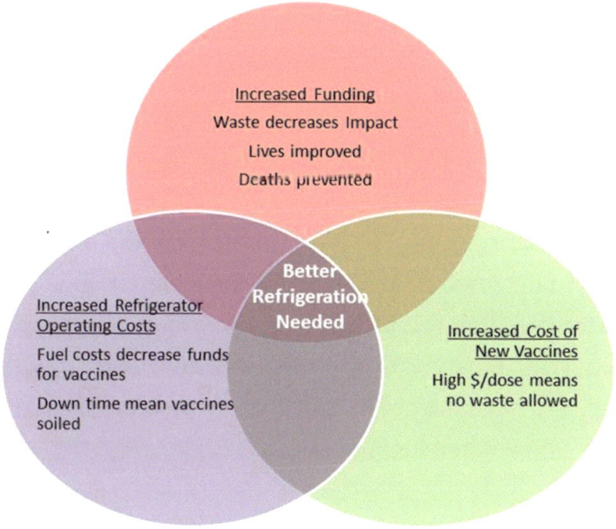

Figure 1: Market forces and their impacts increase the pressure to solve cold chain issues.

temperature between 4°C and 8°C throughout the supply chain of vaccines (called the "cold chain") to remain viable until administered. Vaccine order quantities from public health officials build in an expected loss of 50%. A large portion of this loss is due to exposure to heat or freezing due to improper refrigeration, breakdown of equipment or loss of power. This waste is expensive, accounting for nearly $10 million dollars in Uganda in 2009. (PATH, 2007) Better refrigeration is needed to reduce this waste.

The issue of waste is an overriding impact on each of the market forces. As international funding for immunization has increased, Cool Mile predicts that donors will no longer tolerate half of their money being wasted. For example, The Bill and Melinda Gates Foundation committed $10 billion to immunization for the next 10 years. Their focus is on impact – lives improved and deaths prevented – and they view waste as direct offset to potential impact. As they increase their funding, they are putting pressure on the beneficiaries and the international organization to come up with a solution to the cold chain issues. (PATH, 2008) In the past, the solution has been to decrease dependence on the electrical grid by using alternate fuels like propane and kerosene. These fuels are expensive, diverting funds from purchasing and delivering vaccines, and have their own supply chain problems. Looking ahead to a time when fuel prices could triple, power inputs of this type could consume most of the operations budget immunization budget.[3]

Perhaps the largest impetus for new cold chain solutions comes from the development of new vaccines for diseases such as malaria, roto virus, HIV and dengue fever. The cost of these vaccines will be approximately $7/dose as compared to $0.15 for traditional vaccines measles or DTP. At this price, wasting 50% of these vaccines is unacceptable to donors. (PATH, 2008)

These market forces are putting pressure on countries to purchase reliable, robust refrigerators that can run well on solar power or the existing electric grid, with its frequent volatility. The Outreach Cooler meets the needs of the purchasing customer, the user at the health center, and the funding organization to decreased waste in the cold chain.

B. Succeeding where others have failed

The primary objective of Cool Mile is to increase human wellbeing in low and middle-income countries by increasing immunization through providing robust, reliable, mixed-energy input refrigeration products. The Outreach Cooler is positioned as the solution for vaccine storage in rural health centers where electricity is unreliable. The product can also be used for transporting vaccines during immunization

---

[3] This is a Cool Mile estimate based on the WHO Comprehensive multiyear plans that include detailed costing for purchasing cold chain supply as well as health center operating costs and the purchase of vaccines.

campaigns where healthcare workers go out into the field to administer vaccines. Secondary markets include storage and transportation of any type of medical tissue and blood samples, vaccines and medicines for livestock and domestic animals and medicine storage and transport for disaster relief missions and military operations. The secondary markets are not addressed in this marketing plan for the launch of the Outreach Cooler. The Cool Mile project may broaden its focus after market penetration in the initial market of vaccine storage and transportation has been achieved.

Customer need - Increased Vaccine Coverage: The target customers are the purchasers of vaccine quality refrigerators. They are looking for refrigerators that have a long life, require minimal maintenance, can run on a fluctuating power supply and/or on solar power and have a lower overall "whole life-cost" than the units purchased today. The users will be healthcare workers in the clinics. They require a low maintenance refrigerator that works, enabling them to care for the people in their area. Both have the goal of increasing the number of people vaccinated, especially children, in their catchment area and in their countries.

Over 3 billion people live in rural areas in low and middle-income countries. About one half of the health centers serving these people are connected to "dirty grid" and the remaining are without electricity. (PATH, 2005, 2009; Village Reach, 2008) A rural health center is typically situated where travel is difficult due to roads and infrastructure and the supply of electricity is characterized by frequent "brownouts" and "blackouts". (Ho, Owusu, Aoki, 2009) Vaccine refrigeration in these areas typically has a back-up energy source of kerosene or propane, which is expensive and has supply issues of its own. Health centers without electricity run equipment solely on either kerosene, propane or solar. Breakdowns, lack of spare parts, maintenance and fuel delivery are formidable issues for these centers.

Figure 1: Health Center in Nagaseke, Uganda, 33 miles from the capital of Kampala. The roads make travel difficult and a generator powers the center. Source: Melissa R. Ho, University of California, Berkeley

Purchasers of vaccine refrigeration for rural health centers fall into three segments – 1) country-specific Ministries of Health (MOHs) that allocate funds for cold chain equipment purchases; 2) Public and private health centers that purchase equipment directly from suppliers and 3) donor and country level agencies that give aid to low and middle income countries for cold chain purchases.

	Ministries of Health (MOH)	Public and Private Health Centers	NGOs and Donor Organizations
Customer Segment Overview	Direct Purchaser	Direct Purchaser	Primarily Indirect – Provides funds and direct donations of equipment
Cash Source	WHO, UNICEF, Government (e.g. USAID), GAVI, Internal Revenue	Revenues from fees for service, donations	Donations, United Nations Funds
Procurement process	Through UNICEF catalog (30% of purchases) or Agent (e.g. Afritool)	Direct purchase or agent	
Market Size	37 Countries in Sub Saharan Africa		

2011 est spending on Cold Chain Equipment	$23 M

V. The Competition

A. Current Competitors

There are currently eight vaccine quality refrigerators on the WHO PQS qualified list. These eight products come from the three manufacturers, the largest being Dometic. All of these suppliers are privately held businesses and it is difficult to find market share or sales information. Estimates by PATH show Dometic having the largest market share of vaccine coolers for low and middle-income countries. Each of these competitors has a line of household and recreational refrigeration products and the vaccine and medical units are a division within their organizations. As such, their medical product lines are modifications of their other existing household units. Based on an inventory audit by PATH in 2007, the refrigeration units available today are larger than what is needed for most rural health centers. Running a refrigerator that is larger than needed is costly and creates the incentive to turn it on and off when not in use leading to breakdowns.

Three of the eight units on the WHO PQS list are solar units. From this fact and discussions with and other cold chain experts, it is Cool Mile's impression that ministries of health and donors are interested in moving towards a preference for purchasing solar powered units over kerosene and propane models. The timing is right to enter the market with an efficient, cost effective, solar powered vaccine cooler.

Refrigerator Type	Strengths	Weaknesses	Competitive Advantage
Electric/Propane Electric/Kerosene (Dometic XXX;	• Mixed energy input • Available now • Sales and distribution in place now • Experienced selling into this market and in Africa	• Whole life costs are high (1/2 of the cost is due to energy) • Needs a high level of energy to start. • Degrades quickly under "dirty" grid energy input • Depends on gas distribution • Avg. life = 10 years • Working fluids are often hydro-fluorocarbons	• FSPC uses solar and electricity eliminating the cost of fuel • Significantly lower whole life costs • FSPC runs well in dirty grid • Manages fluctuating energy inputs well without degradation. • Avg. life = 30 years • FPSCs use no CFC or other environmentally dubious chemicals. Uses the environmentally safe, inert gas helium as the working fluid • Lower maintenance due to fewer moving parts

Refrigerator Type	Strengths	Weaknesses	Competitive Advantage
		(HCFCs), which are harmful to the environment Kerosene fridges are unreliable (PATH)	
Ice Lined refer	• Low cost and available anywhere	• Risks freezing vaccines due to the variable temp inside. • Environmental concerns (HCFCs)	• Same as above plus no risk of freezing

Refrigerator Type	Strengths	Weaknesses	Competitive Advantage
		• Electricity only and does not function well in dirty grid • High risk for expensive vaccines	
Solar with batteries	• Long holdover time • Off the grid • Vaccine quality	• Unsuitable for places w/o long solar days • Batteries are expensive and prone to "flooding"	• The Stirling engine runs more efficiently on solar and other combustion engines and can use more of the solar day.
Solar without batteries	• Off the grid Low Whole life costs	• Hold over time as unproven	• Same as above

Refrig erator Type	Strengths	Weaknesses	Competitive Advantage
	• Vaccine quality		

(The above competitive comparison comes from documents created by PATH, a 3rd party with no vested interest in any product, the WHO PQS data sheets and competitor websites. See Appendix)

B. Potential competitors

There are other refrigeration manufacturers that might enter the market. These include research and biomedical refrigeration companies whose products are used for collection of samples such as tissue and blood and or short-term storage. The companies that Cool Mile knows of that have these products, have not made any efforts to have their products PQS certified. The biggest unknown threat is a radical breakthrough in technology based on biomimicry that eliminates the need for cooling of vaccines altogether. This technology is not in the trial stage as of yet which means it will not be ready for commercial use for at least 10 years.

Potential Competito	Strengths	Weaknesses	How you are better

r			
Powder form vaccines that don't need refrigerati on	This is the Holy Grail of the immunization community. This would remove the need for refrigeration all together.	There is some testing on this but it is not in trials and it is not clear which vaccines it would be appropriate for. Cost could be an issue as syringes currently cost $0.08	Refrigerati on is available now.

C. Competitive Strategy

The competition in the market is made up of long time players in the international refrigeration market. Dometic is the market leader with 50% of the market for vaccine and medical refrigeration. The Outreach Cooler competes directly with other solar powered refrigerators and as a substitute for electric/gas and electric/kerosene models that have dominated the market over the years. It is these products however, that are contributing to the wasted vaccines. Reliability is a problem and fuel and efficiency are concerns. Customers will choose Cool Mile's product due to lower whole life costs, long life, reliability and ease of use and maintenance. The Stirling engine technology gives the Outreach Cooler an advantage in removing some of the reliance on humans to maintain the refrigerator and monitor the temperatures by providing robust engines that thrive in conditions found in rural settings.

VI. The marketing and sales plan

A. Marketing Communications Plan

Message

The Outreach Cooler safeguards vaccines and expensive medicines in a portable, lightweight, solar powered refrigerator that runs well in tough environments. It is the best refrigerator for vaccines and medical needs because it will run for 30 years, uses less energy through solar or electricity, and requires little maintenance. The "whole life" cost is much lower than existing refrigerators. You can count on the Outreach Cooler to keep your investment safe.

Media

Cool Mile will reach their customers through an integrated marketing plan. The goal of these marketing efforts is to create awareness and demand for the Outreach Cooler. The basic elements will be:

- A website detailing the results of the in-country trials and customer testimonials
- Data sheets with customer benefits
- Technical specifications sheets
- Placement on the WHO PQS list and UNICEF procurement catalog
- A decision tree on how to choose a refrigerator that is right for your situation
- A "whole-life cost" comparison

Awareness of The Outreach Cooler product will be created by demonstrations at donors and International NGOs headquarters and at conferences and events centered in global health. In addition, PR will be generated about 1) how the product helps increase immunization 2) the technical advances and a YouTube video highlighting customer benefits. Consideration will be created by an installation training video on YouTube that shows how easy it is to install, maintain and monitor. Purchase will be available through UNICEF and in country agents. Price will be competitive with other available units. Service and support will be through maintenance contracts.

Cost

The marketing activities are mostly demonstrations, sales calls, data sheets and word of mouth with no costly advertising. Sales, General and Administrative costs will run $350k per year.

B. Business Model and Pricing

Business Model

This is a traditional business-to-business retail sale model where the Cool Mile division contracts with TwinBird to manufacture Outreach Coolers to the stated specifications. Cool Mile handles marketing, driving awareness and demand at the donor and NGO level through presentations, demos and conference participation. Donors and NGO have indirect influence over the purchase decisions of Ministries of Health. A Cool Mile salesperson will call on purchasers like Ministries of Health and private hospitals in target countries. In addition, Cool Mile will contract with a local agent to handle distribution and some sales. At the end of life of the product or after 10 years, the purchaser will be able to upgrade to the newest product from Cool Mile for a 10% discount of the original product is returned for proper disposal and/or reuse of parts.

Pricing

The retail price of the unit will be $2000 USD without solar panels. This unit can work with either solar panels or electricity from the grid. Maintenance contracts can also be purchased. This price is similar to other products of the same size and configuration. There is a 50% gross margin on the units purchased. (see Summary Financials)

Service

Monthly maintenance is simple and easy. There is a filter on the outside that can be cleaned by removing lint and dirt. Annual service and repair due to accidents can be purchased from Cool Mile. It is beneficial to the customer to have trained technicians to service the units periodically. Service contracts also enable Cool Mile to ensure customer satisfaction, get feedback on the product from customers while in use on site and to influence the follow on purchases of cooling equipment. The service contract concept will be tested with customers during The Outreach Cooler's in country test. It is not currently included as a revenue stream or cost in the financials due to conflicting advice about whether customers will buy a service plan. It has been agreed to test the concept with customers before it is built into the business plan.

C. Marketing Sustainability

Cool Mile's customers are concerned with the wellbeing of their citizens and the communities in which they serve. Their goals are aligned with Cool Mile's in the desire to reach as many people as possible with vaccines. Cool Mile's marketing will focus on how their product will enable them to do that in a portable, cost efficient fashion. The light weight, reliability, long life and flexible power inputs of the product are key benefits for the buyer, reducing the need to replace the unit as frequently or source propane and kerosene as power source.

Over and above the usage benefits of their product, Cool Mile will promote the environmental aspects – no CFCs or HCFCs - and the ability to run on lower electricity and on solar. Cool Mile has a clear competitive advantage in the energy used and in removing a dangerous refrigerant like HCFCs from a refrigerator. The gas used for cooling in the product is helium, which is a neutral, non-toxic gas with no biological role. Toxicity after the product is thrown away or abandoned is a primary concern of WHO and other major donors, when supporting cold chain products. Cool Mile donor demos and marketing will highlight these technological advantages.

VII. Operations Plan

A. Introduction

Global Cooling Inc will create a new division of their company specifically for the Cool Mile product line. The Cool Mile division will be responsible for the following operations:

- Product Requirements to meet market needs
- Sales & Marketing - including
    - Marketing to build awareness and demand
    - Business Development to establish customer relationships, sales channels and customer service
    - Sales Order Processing

Manufacturing of the final end product will be the responsibility of Global Cooling's other operating divisions or contract manufacturers (e.g. Twinbird) and other material suppliers (e.g. solar panels, phase change materials, etc.).

The remainder of this document will outline the key processes and details around Cool Mile's key operations.

B. Product Development and Launch Plan

The development of a prototype that meets PQS requirements is underway. An initial concept was used in a recent field test by PATH in Vietnam. The results from that test look positive and the next phase of development has begun. The product is on schedule to be ready for an in country test with users in the fall. The overall launch schedule is below. The target launch date is December 2010 and the product will be demonstrated at the Geneva Health Forum next May. Details of the launch plan can be found in the Appendix.

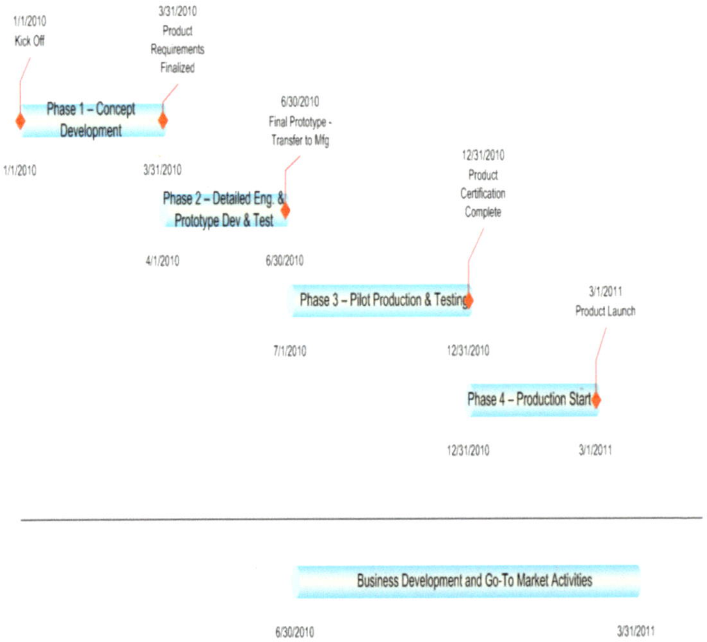

C. Business Development and Sales Operations

Business Development Operations

Launching Cool Mile vaccine refrigerators for use in the cold chain will require a holistic sales and business development effort as the players in the movement toward greater global health act as an interdependent ecosystem. Since 2000, donations for immunization have increased, triggering a scale-up of investments in cold chain and other capital equipment. An analysis of 50 low-income countries that received GAVI support for new vaccines showed that the first year of receiving GAVI funding, total expenditure on cold chain equipment and maintenance increased by an average of US$1.86 million.

Donor organizations and NGOs are highly influential in what gets purchased. If influencers like the Bill and Melinda Gates Foundation and the Clinton Foundation are enthusiastic about the Outreach Cooler, Ministries of Health will pay attention and include the Outreach Cooler product in their consideration set. In addition, researchers at PATH believe that, as more expensive vaccines are brought to market over time, donor influence will increase and they may purchase refrigeration equipment along with the purchase of vaccines.

Cool Mile's business development strategy will involve relationship building amongst donors, NGOs, ministries of health, and agents through product demonstrations, donor outreach, and conferences. Seeing the product is an essential part of the trial and purchase process of cold chain equipment and most of Cool Mile's sales and marketing efforts will be spent on demos and conferences. Some conferences Cool Mile will attend include:

- 4th Vaccine and ISV Annual Global Congress, October 3-5th, Vienna, Austria
- Geneva Health Forum, May 25 – 28th, Geneva, Switzerland
- East, Central, and Southern African Health Community – Health Ministers Conference (location alternates annually between member countries)

Business Development Staffing

Cool Mile's business development team will consist of two employees during 2011-2012. One based in the United States to work in close proximity with the Project Optimize, PATH, the Gates Foundation, the Clinton Foundation, and GAVI's US office.[4]

The second business development person will be based out of Europe to focus on developing relationships in Africa with Ministries of Health in Cool Mile's targeted countries, regional UNICEF offices, and agents. A European office will allow us to develop closer relationships with WHO, GAVI and UNICEF whose headquarters are located in Europe.

	2011	2012	2013	2014	2015	*CAGR*
SG&A*	$340	$360	$ 470	$ 470	$ 480	*6.1%*
% of sales	*25.8%*	*18.4%*	*20.8%*	*16.7%*	*14.0%*	
% growth Y-o-Y		*0.6%*	*30.2%*	*1.3%*	*1.3%*	

- Numbers in Thousands of Dollars. Please refer to the Cool Mile 5 Year Financial Model for detailed SG&A costs.

Direct Sales Operations

All orders placed through the Direct Sales Channel and in-country distributor network will work directly with the Cool

---

[4] GAVI (Global Alliance for Vaccines and Immunization) is the central international organization focused on Immunization.

Mile Sales Team. Cool Mile will work closely with the MoH to determine the correct unit and corresponding solar panel to support the health center's needs. The MoH will place orders with Cool Mile or the distributor directly depending on the relationships established in that region. All direct orders will utilize the logistics and shipping relationships of the agent in each individual country. The transfer of liability will take place at the port of entry in the respective country. The agent will then be responsible for final delivery of the products to the MoH's national store or if requested, to the health center.

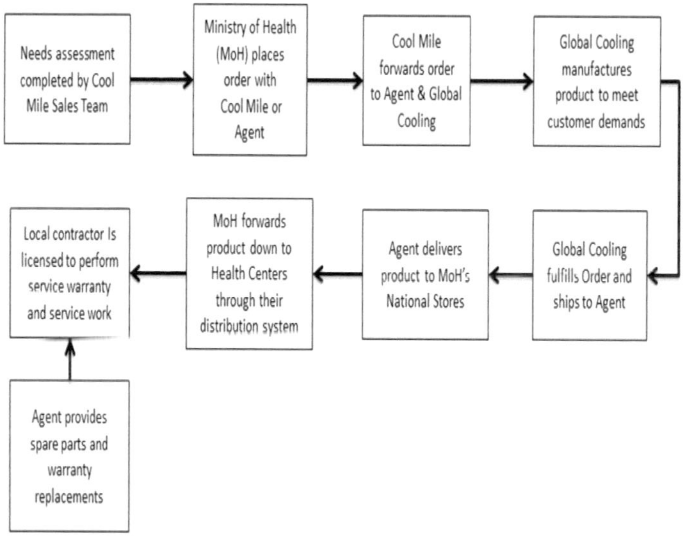

Indirect Sales Operations

Orders placed through the indirect sales channel will work directly with UNICEF's Procurement Service and the UNICEF country office in that region. In this instance, it would be UNICEF that is working closely with the Ministry of Health to determine the correct cooler and corresponding solar array that would meet their needs. The Cool Mile Sales Team will work with both parties to ensure that the products ordered will be suitable for the customer's refrigeration needs.

All orders through UNICEF will be shipped directly to UNICEF's Procurement Service (PS) in Copenhagen. The PS will then be responsible for logistics and shipping product is delivered to the Ministry of Health's National Store.

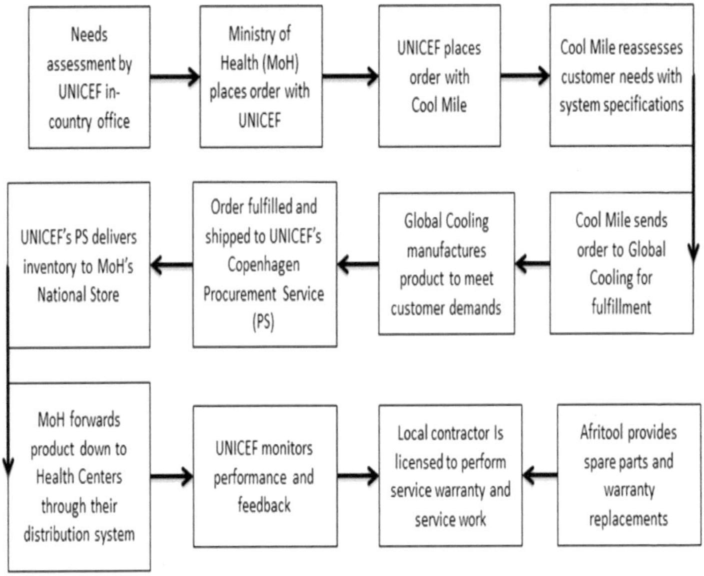

Training, Service & Warranty

The Cool Mile Sales Team will offer training to Ministries of Health staff members and local agents to train staff on the operations and maintenance of their products. All operations manuals will be available in multiple languages on the website to ensure proper operation and service. A sticker with daily operations will be affixed to the unit itself.

Global Cooling offers a 5-year warranty on parts and a 1-year warranty on labor. Spare parts are included with each order of coolers and will be delivered to the National Store for storage. They can also be purchased through the distributor.

Service is contracted in each country through either a licensed contractor or with the Ministry of Health's district repair centers. Due to the limited amount of moving parts in Cool Mile's units, most repairs should be able to be completed at the regional level.

Cool Mile Sales Office

The Cool Mile division works remotely removing the need for overhead expenses for office space. Global Cooling headquarters will supply any necessary office equipment (computers, etc.) as needed by the Cool Mile employees. If it is determined that physical offices are needed in the future, the organization is open to considering options at that time.

D. Supply Chain Management

Manufacturing of the final product will either be located at TwinBird (see Partner Profile), a contractor to Global Cooling located in Japan or at the Global Cooling plant in Athens, OH. The location of the manufacturing facility will be determined once the final product is complete and the engineering department finalizes the components required to pass PQS pre-qualification. A small amount of inventory will be held at the manufacturer and orders will be shipped directly to the customer. The customer will pay for shipping. Cool Mile will hold no inventory aside from demonstration products.

Global Cooling

Global Cooling's manufacturing divisions will handle supply chain relationships, logistics and procurement. This includes managing the orders for final products, if contracted out, solar solution components, etc.

The actual ordering and procurement process will vary depending on the sales channel through which the order was placed (directly or indirectly with an agent, UNICEF, the website). Global Cooling will manufacture and ship product to order, as neither channel will warehouse an inventory of Global Cooling's products.

E. Operations Sustainability

The Cool Mile Business unit is a profit center within Global Cooling. The Outreach Cooler is purchased from the contracted manufacturer and is a modification of an existing product. The division sells, markets and distributes the units. As such, Cool Mile's first focus will be on the downstream operations of fair pricing, fair treatment of all employees and contractors and transparency with stakeholders, especially donor organizations. TwinBird in Japan controls the inputs into the construction of the Outreach Cooler product and Cool Mile will work with them to use as many sustainable and environmentally friendly materials as possible. Cool Mile is investigating their practices more closely and understanding under what regulations they operate, especially with regard to the Kyoto Protocol.

VIII. Venture Team

Daniel Byrne, Founder & Business Development co-lead

Daniel Byrne is a proven entrepreneur and leader. After formal training in engineering and sales experience with two major corporations, Dan founded Byrne Specialty Gases, Inc. in 1984. The vision was to provide engineered gas solutions, market applications and world-class service to biotech companies, research and development firms, environmental and medical laboratories and electronics manufacturers. By 2005, BSG had become a nationally recognized leader in the life sciences and analytical laboratory fields generating more than $10 million in annual sales. In March of 2006, BSG was acquired by Airgas, the largest U.S. distributor of industrial, medical, and specialty gases.

Craig Nakagawa, Founder & Business Development co-lead

Craig has spent the past nine years building VillageReach into a social venture recognized for its impact in health and development. VillageReach's innovations in Last-Mile health logistics, cold chain, and social business are credited with raising immunization rates from 68.9% to 95.4% in the province of Cabo Delgado between 2003 and 2008 versus the national average of 72%. He has held every executive position, including President, COO, CFO, and Director of Social Business. Craig is particularly proud of his leading roles developing VillageReach's original field data system, the foundation of the current award-winning MIS system, and building the LPG distribution company, VidaGas, into a recognized example of social business innovation.

Lara Stack – General Manager

Lara has 10 years of consulting experience assisting Fortune 500 companies with large-scale technology implementations, security and controls consulting, business analysis, process improvement and project management. Lara is currently a Senior Product Manager at T-Mobile USA where she manages a portfolio of handsets and is responsible for managing go-to market activities across the T-Mobile enterprise and its channel partners. She also has a passion for sustainability in business and is pursuing an MBA at the Bainbridge Graduate Institute (BGI),

Lara has a B.S. in Accounting and Management Information Systems from Purdue University and is also a Six Sigma Green Belt and a Certified Information Systems Security Professional.

Libby Johnson McKee – Marketing Manager

Libby is an MBA candidate at Bainbridge Graduate Institute and is dedicated to empowering people through technology and access to opportunity. She spent most of her business career in game changing businesses that created new ways of thinking, selling and driving to put the customer first. She worked in Product Management and Marketing at Microsoft from 1989 to 1996 and then at Amazon.com in Books, General Management from 1997 to 2001. Since 2001, Libby has managed her own marketing consulting business and served as Managing Director of Agua Limpia, an environmental education nonprofit.

Denny Pallenberg – Operations Manager

Denny is the Sales Manager at To-Go Ware, a green company that makes sustainable products for eating and carrying food on the go. He is also provides support in the areas of marketing and operations and is an MBA candidate at Bainbridge Graduate Institute.

Before To-Go Ware, Denny owned and operated a wholesale spa and home products distribution company. Denny also worked in the transition management division at Barclays Global Investors and pioneered a paperless office plan within that division. He has helped to create BGI's Corporate

Tutoring Program in association with the Boys and Girls Club of America.

Manan Shukla – Finance Manager

Manan is employed as Sr. Analyst for Primus International, a Bellevue-based global aerospace manufacturer owned by Oak Hill Capital, a private-equity firm based in New York. His current responsibilities include: strategic and financial planning, modeling, analysis, project management, and systems implementation.

Prior to Primus, Manan spent four years with Booz Allen Hamilton where he worked on multiple engagements providing financial management solutions to an array of military clients, ranging from the US Navy to the Air Force.

Manan graduated from Lehigh University in 2004 with a BS in Finance. He is currently pursuing an MBA in Sustainable Business at Bainbridge Graduate Institute and will graduate in June '10.

IX. Partnering Strategy

Cool Mile division will be partnering with several organizations to ensure the successful development and launch of the Outreach Cooler. Cool Mile's partners include the following groups and organizations:

Global Cooling (www.gloablcooling.com)

Global Cooling has developed unique, green technology that employs a natural refrigerant, helium, to provide cost effective, environmentally friendly, long lasting cooling equipment. Their products consume significantly less energy than conventional technologies and help their customers cut their energy costs by more than 50% and reduce their carbon footprint.

Global Cooling offers low energy consumption, lightweight, portable freezers and ultra-low temperature freezers. Global Cooling also provides OEM customers with their oil free, zero maintenance Stirling cooling technology for inclusion into their products.

PATH (www.path.org)

PATH is an international nonprofit organization that brings together public and private sector partners to develop health technologies and strengthen health systems to deal with the world's pressing health problems. PATH bridges the gap to bring technologies to market before they are economically viable for commercial ventures to enter the market in a profitable manner.

Cool Mile works closely with PATH's technical experts and commercialization specialists to find the ideal markets and countries to test and expand distribution of their refrigerators.

TwinBird Corporation (www.twinbird.jp)

The **TwinBird** Corporation is Global Cooling's contract manufacturer, which currently manufacturers the chassis and final production of the SC-BV25 cooler. TwinBird Corp. acquired ISO9001 certification on 1996 and ISO14001 that assures the product quality and environmental standards of their products. TwinBird is location in Tsubame City, Japan. With over 50 years of manufacturing experience their manufacturing processes specialize in household appliances such as: vacuum cleaners, waterproof audio-visual appliances, lighting fixtures, home consumer appliances, and other refrigeration products.

Afritool (http://www.afritool.com/profile.html)

Cool Mile is in the process of developing a dealer relationship with **Afritool**, a distributor with over 12 years of experience working in the cold chain equipment market. Their experience with cold chain equipment and developed logistics operations will provide Cool Mile with an effective partner for quality and on-time delivery of their products.

Other Key Suppliers

Cool Mile is currently in the process of determining the solar and phase change material suppliers that will be critical in the final production of the Outreach Cooler.

Other NGO Partners

Cool Mile's business development team will continue to pursue opportunities and partnerships with other NGO organizations that have the ability to influence the purchase decision of vaccine coolers at ministries of health, as well as in bilateral and unilateral donations of cold chain equipment. Other NGO organizations that will be targeted include The Bill and Melinda Gates Foundation and The Clinton Foundation.

X. Key Risks and Assumptions

Assumption #1: There is an increased need for cold chain refrigeration in sub-Saharan Africa due the following factors:

- An increase in new vaccines coming available over the 3 to 15 years.
- Population growth in this region
- An increase of storage space required due to new single dose packaging requirements for the more expensive vaccines

Associated Risks:

- Timeline of new vaccines coming on the market and available to developing countries is not guaranteed
- Alternate or breakthrough solutions to vaccine delivery will become available

minimizing the need for increased refrigeration in this market

Assumption #2: Last mile health clinics and outreach centers desire the following characteristics in its vaccine refrigeration products:

- Stable temperature control
- Solar energy input option to replace kerosene and propane solutions
- Lower maintenance and longer lasting product
- Durable and portable solution

Associated Risks:

- Ministries of Health, Bi-laterals and Multi-Laterals are unwilling to pay for solar solution due to risks of theft and cost of set up and panels
- Longer life and reliability of product may not persuade customers to pay a premium for the Outreach Cooler product solution

Assumption #3: The Outreach Cooler product is capable of meeting and passing the standards required for the World Health Organization's PQS standards.

Associated Risks:

- The Outreach Cooler product does not meet the needs for PQS qualification and therefore is not listed on the World Health Organization site for approved cold chain equipment.

- Not meeting these standards will not allow for The Outreach Cooler product to be listed on UNICEF's equipment procurement catalog

Assumption #4: Final product price is comparable to other solutions available in the market, yet more costly to produce.

Associated Risks:

- Final product solution requires a higher price than other similar sized products available in the market

Mitigation Plan

Cool Mile recognizes that there are several risks with pursuing only the market in sub-Saharan Africa. It will be imperative for Cool Mile and its parent company to also pursue other opportunities in other markets such as disaster relief, blood, bone marrow, and organ donor storage and transport. In conjunction with pursuing these other markets, Cool Mile will continue to monitor and work with PATH and other NGOs to assess the vaccine development and how it will affect the cold chain market in sub-Saharan Africa and therefore the demand for Cool Mile's Outreach Cooler.

The Outreach Cooler is based upon an existing product and configuration that is already being produced by Global Cooling's partner TwinBird Manufacturing. Current Global

Cooling and PATH engineers are working together to tweak and test the current design in order to meet PQS standards. While there is still a small risk that the product may not pass, Cool Mile is testing all aspects of the product against the PQS standards prior to submitting the final product for validation and testing at the World Health Organization. In addition, since the outer casing and the majority of the components will remain unchanged, it is expected the ramp up manufacturing costs (e.g. tooling) costs will be minimized allowing us to keep the product in the price range of $1500 to $2000 that is price comparable against other products of similar size and function in the market today.

As Cool Mile is targeting a price point that is in line with other competitor products, their sales and marketing strategy will bring more awareness of the benefits of the Outreach Cooler product over their competition eliminating a key portion of the risks identified above. As Cool Mile gains more insight into their market they will continue to seek solutions to address some of the other risks, such as solar panel theft.

XI. Financials

Pro Forma(in'000s)	2011	2012	2013	2014	2015	CAGR
Revenue	**$1,380**	**$1,949**	**$2,245**	**$2,829**	**$3,421**	*19.9%*
% growth Y-o-Y		*41.2%*	*15.2%*	*26.0%*	*20.9%*	

Gross Margin	**$ 644**	**$ 909**	**$ 1,048**	**$ 1,320**	**$ 1,597**	
% of sales	*46.7%*	*46.7%*	*46.7%*	*46.7%*	*46.7%*	
SG&A	**$ 356**	**$ 358**	**$ 466**	**$ 473**	**$ 479**	*6.1%*
% of sales	*25.8%*	*18.4%*	*20.8%*	*16.7%*	*14.0%*	
% growth Y-o-Y		*0.6%*	*30.2%*	*1.3%*	*1.3%*	
EBITDA	**$ 287**	**$ 551**	**$ 581**	**$ 847**	**$ 1,117**	*31.2%*
% of sales	*20.9%*	*28.3%*	*25.9%*	*30.0%*	*32.7%*	
% growth Y-o-Y		*91.5%*	*5.5%*	*45.8%*	*31.8%*	
Selling price (USD)	2011	2012	2013	2014	2015	-
Base Unit	$ 1,362	$ 1,362	$1,362	$1,362	$1,362	
Base + Solar Unit	$ 2,000	$2,000	$2,000	$2,000	$2,000	
Total Units Sold	2011	2012	2013	2014	2015	
Base Unit	229	323	373	469	568	
Base + Solar Unit	534	754	869	1,095	1,324	
Total Units Sold	**763**	**1,078**	**1,242**	**1,565**	**1,892**	

XII. Summary

The Outreach Cooler is the initial product from the Cool Mile division targeted to using Stirling Engine technology to increase health and wellbeing by increasing access to immunization. The business is breakeven in year two and an NVP of XXX. The mission of the division is to pioneer vaccine and medical cooling technology that can be sold or licensed to other refrigerator manufacturers .

Messages – breakeven to profits. TBL, social good, great partners that want this to happen capitalize on market forces.

XIII. Appendix

A. Global Cooling Organization

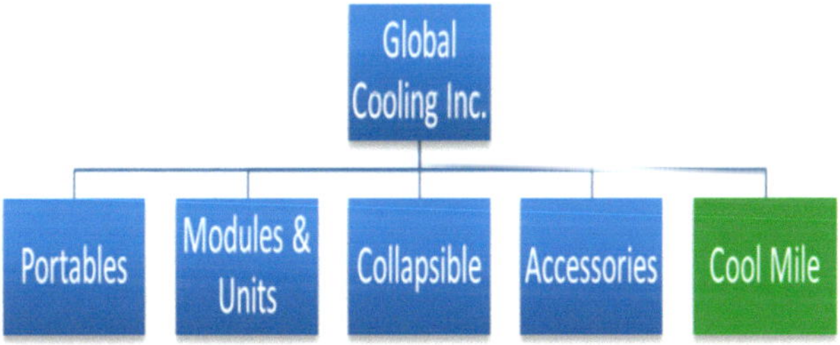

B. Market Intelligence

Customer Segmentation

Ministries of Health: The role of the MOH in a country is "...to promote equity in health, to combat disease, and to improve the quality of, and lengthen, the lives of the peoples of that nation." In that capacity, these governmental organizations are responsible for the health systems in their countries ranging from health policy and prioritization to delivery methods and personnel. Ministries of Health are the primary purchasers or sources of funds for purchases of cold chain equipment. The sources of funds for these purchases from International NGO's like World Health Organization (WHO), UNICEF, The Global Alliance for Vaccine and Immunization (GAVI), individual country organizations like USAID and others. Typically, a MOH will develop a procurement plan for vaccines and equipment and apply for funds from these organizations to support it.

Once the funds are received, the MOH can purchase directly from the distributor or manufacturer or through UNICEF and WHO, which certifies equipment for vaccine storage. While only 30% of refrigerator purchases are made through UNICEF, it will be critical for the Outreach Cooler to be WHO certified and on their supply list. This will provide confidence in the product, enable direct-from-distributor purchases and, equally importantly, will enable donors and other funders to recommend the Outreach Cooler unit.

There is one MOH per country and 116 countries that qualify as low to middle income according to the World Bank (2008), 55 of which are in Africa. GAVI has 72 approved countries that are predominately low-income nations. Cool Mile's first target will be GAVI countries in Africa (37) due their need (low income) and their access to funds for cold chain equipment from GAVI.

Health Centers: In some countries, public health centers determine their needs and buy equipment directly. The funds may come from their own fundraising or from the MOH. Each center can determine the brand and type of unit to buy. Cool Mile believes that this practice is a small percentage of the purchases and that the arrival of more expensive vaccines will reduce the incidents of independent purchases. The MOH will want to have more control when the cost of waste is high. Cool Mile estimates that, in low and middle-income countries, there is one health center for every 7,800 people, which equates to about 700,000 centers. In Africa the ratio is higher at about 10,000 people per center or approximately 100,000 centers. (2005, World Bank)

Population growth around the world is concentrated in less developed countries. Of the world's 6.8 billion people, 5.5 billion of them are in less developed countries. It is expected that the population of Africa will double to 2 billion people by 2050 and that nine of every ten youths will live in a less developed country. (Population Reference Bureau, 2009) These statistics support Cool Mile's assumption that the number of health centers must grow to support the increasing number of children. The volume of vaccines will continue to grow to match the population and supportsthe rationale for increased funding for immunization from the international donor organizations.

NGO's and Donor Organizations: International organizations are mobilized to increase global health. The announcement by The Gates Foundation of their commitment to this initiative in 2001 increased the focus on and awareness of health issues in developing countries. The donors have historically funded research, development and purchase of vaccines. In the last 5 years, they have become aware that a major barrier to their goal of immunization and disease prevention is the cold chain (storage and delivery of vaccines). Recently, they have begun funding equipment purchases through GAVI or donating units directly. For example: The country of Demark purchased 300 refrigerators and gave them to African nations outside of existing procurement channels.

Cool Mile sees donor support and endorsement as critical to seed the *Outreach Cooler* as the preferred refrigerator for the next generation of expensive vaccines. Cool Mile is targeting GAVI, Gates Foundation, PATH and others to recommend the purchase of the Outreach Cooler product to secure their investment in vaccines. This will shift the thinking of refrigerator purchases as independent from the vaccine purchase. In other words, Cool Mile's marketing goal is to link the investment in new vaccines to a complimentary investment in new refrigeration technology. Donors also have influence at UNICEF, WHO and other governmental organizations and their recommendation will help increase awareness of the Outreach Cooler product.

Structure and size of the target market

Rural health centers worldwide are expected to receive $70M in vaccine refrigerators in 2011, according to PATH, the leading nonprofit organization that studies health technology issues in developing countries. Cool Mile's initial target market is Sub Saharan Africa, which is estimated to purchase $23 million in vaccine refrigerators. This is based on historical purchases and the comprehensive multiyear plans submitted by Ministries of Health in 2006.

The target market for the Outreach Cooler is Sub Saharan Africa due to the rapid growth in population. World Bank demographers expect that the population will reach 2 billion by 2050 and over half of these people will be children and youths. Children are the primary recipients of immunizations and, paired with poor electrical grid in much of the continent,

Sub Saharan Africa has both the need and the setting that fits the Outreach Cooler. Based on population growth and the increasing availability of vaccines for different diseases, the market for new vaccine refrigerators is between 15,000 and 20,000 units annually.

C. Product Development

Product Development Team, Roles and Responsibilities

The product development team will consist of engineers from PATH and Global Cooling. They are responsible for the development and testing of the final product specifications and prototype. TwinBird, Global Cooling's contract manufacturer of the SC-BV25 will also be included in this process as product specification details are finalized to determine feasibility of final development.

Product Differentiation

Cool Mile's first product offering is in co-development with its partner PATH and their Project Optimize team. Project Optimize is responsible for assessing the vaccine cold chain system in order to make recommendations for improvements as new vaccines become available over the next five to ten years. As part of their assessment it was determined there is a strong need for improved refrigeration at peripheral health centers in low and middle-income countries across the globe where energy resources are unreliable (dirty grid) or unavailable (off grid).

Given this need, there is a real opportunity for combined use of PV panels and Stirling coolers. Global Cooling and their contract manufacturer TwinBird have a cooler that is already compatible with solar power as it runs on direct current (DC). Another advantage of the FPSC technology compared to standard compressors is its ability to run on low power input. Traditional compressors require a relatively high minimum start up current in order to begin cooling. FPSC, on the other hand, will begin cooling with a very small incoming current. In a solar application this means that more of the energy produced during the day will be usable (it will start cooling earlier in the morning and continue later into the evening). This may also translate into a smaller solar panel size requirement thus lowering the cost of the system. The straightforward application of PV energy generation combined with Stirling Cooler engines makes for an efficient and environmentally benign technological solution to the challenge of off-grid vaccine storage.

The proposed system will take advantage of an improvement in the refrigeration storage box using phase change material (PCM). Traditionally the holdover time – or time that a refrigerator can maintain cold temperatures without any electrical input – for developing-country vaccine refrigerators has been a concern because of the relatively high ambient temperatures involved. According to recent research at PATH, by adding a PCM liner to the refrigerator box it is anticipated that a significant lengthening of the hold over “cold window” can be achieved, making the overall system more efficient and correspondingly decreasing the energy requirements of the system.

The proposed product system will have a PV module on top of the clinic that produces electric current when the sun is shining. This will supply electricity to the Stirling engine attached to the refrigerator. The system will be designed appropriately to allow for storage of excess energy during the day that will provide cooling overnight when the PV panels are not producing energy.

For this first phase of product development, it was determined to leverage an existing product that is manufactured today in order to minimize the costs around designing and manufacturing a new product from scratch, the SC-BV25. Based on Cool Mile’s own research, as well information provided by PATH, Cool Mile believes this product would be ideal as Cool Mile’s initial product offering given the necessary modifications mentioned above and the criteria needed to meet the World Health Organization’s (WHO) Performance, Quality and Safety (PQS) standards and requirements

Product Development Roadmap

The product development plan is laid out in five phases: kickoff and scoping, concept development, detailed engineering, manufacturing transfer, and manufacturing scale-up and ongoing manufacturing and sale. A brief description of the product development plan by phases as follows:

Phase 0 – Complete

Duration: 1 month, January 2010

This phase includes the creation of a detailed product specification requirements document, the development and finalization of collaboration and partnership agreements and the kick-off of the project through a centrally located meeting.

Phase 1 – In Progress

Duration: 3 months, January 2010 thru April 2010

The first product development phase of the project involves concerted research into the possible configurations that the product system may ultimately take. The activities pursued during this phase involve in depth technical research, brainstorming of system solutions, rapid development of prototypes to test concepts and evaluate the merits of different approaches, and timely and repeated design reviews to refine and distill the best aspects of the concepts developed. The hallmark of Phase 1 is the iterative process of conceiving ideas, quickly testing them through design review and rapid prototype demonstration, and cycling back to the concept to refine it. Important contributions to this work will be sought through market research and user acceptability studies. This phase concludes with a refined product requirements specification and system concept worthy of advance to the detailed engineering phase.

Phase 2 – Not Started

Duration: 2 months, May 2010 thru June 2010

By late spring 2010, it is anticipated that the project team will have developed a system concept that is ready for full-blown engineering product development. This will include mechanical engineering, thermal systems design, electrical controls, software implementation, and manufacturing analysis. Once again the technique of iterative design, development, and testing followed by a cycle of review and editing will be employed to assure that the development efforts remain focused on workable solutions that meet the requirements laid out at the start of the project. In addition to the detailed engineering activities, a battery of tests will be used to put the product systems through their courses and confirm performance as anticipated. Lastly, Phase 2 will include an initial inquiry into the intellectual property landscape for the product system being developed, and the regulatory and certification requirements of the product. Phase 2 concludes when a particular system has been engineered, prototyped, tested, and confirmed is ready to transfer to manufacturing.

Phase 3 – Not Started

Duration: 5 months, July 2010 thru December 2010

At this point in the project the manufacturing partner, TwinBird Corporation, becomes much more involved through the receipt of a detailed engineering package containing all the specifications for the proposed refrigeration system. Their prior involvement in design reviews and prototype evaluation will ensure that TwinBird engineers are already familiar with the technology. Phase 3 involves the implementation of tooling, fixtures for assembly, and development of the manufacturing line.

Important Note: This initial product design is based on the SC-BV25 product that TwinBird already manufacturers. It is assumed that timing and cost related to these activities to be minimized where possible.

The primary outcome during this phase will be the first-run production or pilot run. The parts produced during pilot production will be used to conduct extensive field testing as well as a battery of laboratory tests meant to simulate the challenging environment into which these FPSC flexible powered refrigeration systems will be placed and used. Another Phase 3 activity is to prepare for the certification of the product according to the international quality standards including UL standards, CE marking, and WHO's PQS system. Phase 3 concludes when the pilot parts have been thoroughly tested, vetted, certified and reviewed for potential refinement.

Phase 4 – Not started

Duration: 3 months, January 2011 thru March 2011

The final phase of the product development effort will be the advancement of the production capabilities at TwinBird and the initial building of raw material stores and an inventory component parts (e.g. PV panels, etc.). Global Cooling's Cool Mile division and TwinBird will direct these activities as they balance the initial production run requirements with their capacity to produce systems over a given lead time. At the same time, work will be done on further refining the market introduction product launch plans. The product development partners will stay integrally involved as there are invariable issues that arise during the manufacturing scale-up that need to be reviewed in light of the design intent for particular features and attributes. Phase 4 transitions to Phase 5 at the time that factory systems are ready to begin full production and the marketing and product distribution systems are deemed ready to move products out into the marketplace.

Phase 5 – Not Started

Duration: Product Launch, March 1, 2011

This phase represents the continuous efforts required for marketing and post-market surveillance of the final product.

Product Development Start-up Costs

For this initial product offering, the product development costs are estimated to be approximately $920K.

Please refer to the Cool Mile 5 Year Financial Model for additional details.

Ongoing Product Development

Currently the Cool Mile division will launch one product and will continue to work with its partners to identify opportunities for future innovations to add to its suite of Vaccine Refrigerators.

D. Sales Operations

In order to support the launch of the initial product as well as any future products, the Cool Mile line of products and division will have their product section within the existing Global Cooling website (www.gloablcooling.com).

From this site the following information and features will be available:

- Cool Mile Vaccine Refrigerator product overview and specifications
- Support manuals and training information, available in multiple languages
- Customer and Distributor Order Submittal and Processing

The Cool Mile Ordering Portal will be crucial as part of the end customer product offering. It is Cool Mile's intention that the vaccine refrigerators can be modulated specifically around the solar needs of the end customers. Therefore a feature of the website will be to allow customers to input the location of the final destination of the refrigerator in order for Cool Mile to recommend the appropriate solar solution required for that location. This will ensure that the customer is getting the right sized solar panels and is only paying for the parts that they need versus a one size only solution.

The site will also allow for customers to purchase accessories and any necessary spare parts that also may be required.

Distributor information. Cool Mile products will be sold through 3rd party distributors including UNICEF and agents such as Afritool. Information about these and other distributors will also be included in Cool Mile's website.

Sales Channels

Sales of Global Cooling's vaccine refrigerators will mainly be through two channels, direct and indirect sales.

Indirect Sales – United Nations Children's Fund (UNICEF)

UNICEF's programs support community-level services to promote the health and well being of children around the world, including immunization. With a network of offices in 158 countries and territories, UNICEF has the infrastructure to provide advice, procurement, and immunization services for developing nations. Approximately 30% of the cold chain equipment is procured through UNICEF; which represents a considerable and qualified segment of the market Cool Mile must focus on. Cool Miles products must meet the WHO's PQS standards to be listed in their Supply Catalog and procured through the Procurement Service. Once listed, the

PS Centre will manage all distribution and logistics once and order has been placed and delivered to them from Global Cooling or Twinbird's manufacturing facility.

UNICEF's Procurement Service will be an external distribution partner in the indirect sales channel to distribute vaccine refrigerators to Ministries of Health, their regional offices, and onto health clinics.

Procurement Service transactions are often shared between the PS Centre at the UNICEF Supply Division and the UNICEF country office where the refrigerator will be shipped to. Successful entry into this market will require that close relationships be developed with both parts of this distribution channel.

Direct Agent Sales - Countries that choose to purchase Cool Mile's vaccine refrigerators directly will work with Cool Mile's in-country agents, like Afritool. Any entity interested in procuring Cool Mile's vaccine refrigerators will place orders with these agents, who will contact Cool Mile for order fulfillment. Agents will handle all logistics and distribution once products are shipped from Global Cooling's manufacturing facilities (e.g. TwinBird)

Purchasing from a local supplier or agent provides countries with the opportunity to support their local economy and often receive more competitive pricing than through UNICEF. The procurement process can also be much simpler through a domestic supplier versus with a large NGO. WHO prequalification is also necessary when selling through this channel. These suppliers will act as in-country agents for Global Cooling to take orders and distribute products once shipped from the manufacturer.

Cool Mile is in the process of developing a dealer relationship with Afritool, a distributor with over 12 years of experience working in the cold chain equipment market. Their experience with cold chain equipment and developed logistics operations will provide Cool Mile with an effective partner for quality and on-time delivery of Cool Mile's products.

Bibliography

Reports, Websites and URLs

Abt Associates Inc. (2000). Poverty Reduction and Immunizations: Considering Immunizations in the Context of HIPC II Debt Relief. Seattle: The Bill & Melinda Gates Foundation Children's Vaccine Program at PATH.

Alan D. Lopez, C. D. (2006). Global Burden of Disease and Risk Factors. New York: Oxford University Press.

Dean T. Jamison, J. G. (2006). Disease Control Priorities in Developing Countries. New York: Oxford University Press.

Disease Control Priorities Project. (2006). Vaccine-Preventable Diseases. Washington DC: The International Bank for Reconstruction and Development / The World Bank.

Lloyd, J. (2008). SolarChill - Whole Life Cost. Geneva: PATH.

PATH. (2005). PATH Interview Findings; Vaccine Cold Chain Refrigeration Technologies: Assessment of the Public Sector Market. Seattle: PATH.

Sprague, C. (2007). VidaGás: Powering Health Clinics and Households in Mozambique with LPG. United Nations Development Program, Growing Inclusive Markets. New York: UNDP.

UNICEF. (n.d.). Supply Catalogue. Retrieved May 3, 2010, from UNICEF: http://www.supply.unicef.dk/catalogue/

UNICEF Supply Division. (2008). Supply Annual Report. Copenhagen: UNICEF.

Village Reach. (2008). Data Collection and Decision Support Tools to support Rural Health Clinics in Improving Health. Seattle: Village Reach.

Village Reach. (2008). Integrated Support System for Improving Healthcare: "Starting" at the Last Mile. Seattle: Village Reach.

Village Reach. (2009). JSI Work Order: Mothodology for Operations Research: Data Capture, Reporting, and Distribution Models. Seattle: Village Reach.

Village Reach. (2008). Model for Improving Last Mile Healthcare. Seattle: Village Reach.

Village Reach. (2009). Skoll Awards for Social Enterpeneurship Follow on Funding Application, Narrative Section. Seattle: Village Reach.

Widjaya, A., Santoso, W., Moniaga, V., Simbolon, O., Wasito, Suparno, et al. (2006). Evaluation of Alternative Refrigeration Models for Primay Health Centers. Jakarta, Indonesia: Directorate for Epidemiology, Immunization and Matra Health, Director General for Communicable Disease Control and Environmental Health, Ministry of Health Republic of Indonesia, PATH.

World Health Organization. (2005). Manual on the management, maintenance and use of blood cold chain equipment. Geneva: World Health Organization.

World Health Organization. (1998). Safe Vaccine handling, cold chain and immunizations. Geneva: WHO Expanded Program for Immunization.

World Health Organization. (2008). TechNet Consultation 2008. Tunis, Tunisia: World Health Organization.

Bloom, D. E., Canning, D., & Weston, M. (2005). The Value of Vaccination. World Economics, 6(3), 15-39. Retrieved May 3, 2010, from http://www.sabin.org/files/attachment/value_vaccination_bloom_canning_weston.pdf

World Health Organization. (n.d.). World Health Organization: Immunization against diseases of public health importance. Retrieved May 3, 2010, from www.who.int/mediacentre/factsheets/fs288/en/index.html

World Health Organization. (n.d.). World Health Organization: Expected Vaccine Wastage. Retrieved May 3, 2010, from www.who.int/immunization_delivery/vaccine_management_logistics/logistics/expected_wastage/en/index.html

PATH. (2008, June 1). PATH: Publication : Landscape Analysis: Analysis of EVSM Indicators. PATH: A catalyst for global health. Retrieved May 3, 2010, from http://www.path.org/publications/details.php?i=1699

PATH. (2008, June 1). PATH: Publication : Landscape Analysis: Cool Chain Technologies. PATH: A catalyst for global health. Retrieved May 3, 2010, from http://www.path.org/publications/details.php?i=1698

PATH. (2008, July 1). PATH: Publication : Landscape Analysis: Trends in Vaccine Availability and Novel Vaccine Delivery Technologies: 2008 -2025. PATH: A catalyst for global health. Retrieved May 3, 2010, from http://www.path.org/publications/details.php?i=1700

PATH. (2009, February 1). PATH: Publication : Optimizing Vaccine Supply Chains. PATH: A catalyst for global health. Retrieved May 3, 2010, from http://www.path.org/publications/details.php?i=1701

PATH. (2009, October 1). PATH: Publication : Optimize Strategy: 2009-2012. PATH: A catalyst for global health. Retrieved May 3, 2010, from http://www.path.org/publications/details.php?i=1664

Helmenstine, A. M., Ph.D.. (n.d.). 10 Helium Facts - Quick Facts about Helium. *Chemistry - Periodic Table, Chemistry Projects, and Chemistry Homework Help*. Retrieved May 13, 2010, from http://chemistry.about.com/od/elementfacts/a/10-helium-facts.htm

CFCs and Ozone Depletion. (n.d.). *Center for International Earth Science Information Network*. Retrieved May 13, 2010, from http://www.ciesin.org/TG/OZ/cfcozn.html

World Health Organization Fact Sheet: Immunization against diseases of Public Importance- Retrieved May 13, 2010, from http://www.who.int/mediacentre/factsheets/fs288/en/index.html

Melissa R. Ho, Emmanuel K. Owusu, and Paul M. Aoki. HYPERLINK "http://www.melissaho.com/papers/ictd09-mho-stakeholder.pdf" Claim Mobile: Engaging Conflicting Stakeholder Requirements in Healthcare in Uganda . In Proceedings of Information, Communications Technology, and Development 2009 (HYPERLINK "http://www.ictd2009.org" ICTD2009). April 18-19, 2009.

Competition

Solar Vaccine Refrigerator (Fridge) . (n.d.). Bright Light Solar. Retrieved May 3, 2010, from http://www.brightlightsolar.com/technology/solarfridges.html

Solar power for remote locations, vaccine refrigeration, back up power supply - Dulas. (n.d.). Professional Services for Renewable Energy - Dulas. Retrieved May 3, 2010, from http://www.dulas.org.uk/?id=60

Vaccine / Medical - sibir.com. (n.d.). Sibir. Retrieved May 3, 2010, from http://www.sibir.com/english/products/vaccine--medical.aspx

Vaccine Storage Sun Frost Refrigerator Freezer. (n.d.). Refrigerators, home compost bins and sustainable living products - Sun Frost. Retrieved May 3, 2010, from http://www.sunfrost.com/vaccine_refrigerators.html

Dometic Medical Systems. (n.d.). Refrigerators and Freezers. Retrieved May 3, 2010, from www.dometic.com/enlu/Europe/Luxembourg/Medical-Systems/Cold-Chain/Refrigerators/Europe/Luxembourg/Medical-Systems/Cold-Chain/Refrigerators/

Made in the USA
Las Vegas, NV
03 November 2026

57274964R00043